W9-BOB-688

LIFEVIEWS

Published by Creative Education
123 South Broad Street, Mankato, Minnesota 56001
Creative Education is an imprint of The Creative Company

Art direction by Rita Marshall; Production design by The Design Lab
Photographs by CLEO Photography, Dennis Frates, The Image Finders (Jim Baron, Alan Chapman, Mark E. Gibson, Bruce Leighty, Michael Lustbader, William Manning, Mark and Sue Werner, Joanne Williams), JLM Visuals (Burton A. Amundson, J. C. Cokendolpher, Charlie Crangle, Richard P. Jacobs, Breck P. Kent, Lowell R. Loudon, John Minnich, Marypat Zitzer), George Robbins, Tom Stack & Associates (Jeff Foott, Milton Rand, Allen B. Smith, Greg Vaughn)

Library of Congress Cataloging-in-Publication Data

Frahm, Randy.
Lakes / by Randy Frahm.
p. cm. — (LifeViews)
Summary: Presents an overview of lakes, including their formation, life span, inhabitants, uses, and ecology.
ISBN 1-58341-244-1
1. Lakes—Juvenile literature. 2. Lake ecology—Juvenile literature. [1. Lakes. 2. Lake ecology. 3. Ecology.] I. Title. II. Series.
GB1603.8 .F73 2002
551.48'2—dc21 2001047898

First Edition

2 4 6 8 9 7 5 3 1

TIMELESS RESERVOIRS
LAKES
RANDY FRAHM

WHERE LAKES are found, life is found. In North America, small muskrats craft homes out of the **aquatic** plants that grow in shallow water. In Africa, huge hippopotamuses wallow in lakes to cool themselves and to relieve their legs of the strain of supporting their five-ton body weight. In Russia, the only freshwater seals on Earth live and play in Lake Baikal.

Lakes are found throughout the **world**. They hold an estimated 30,000 cubic miles (125,000 cu km) of water. Almost 70 percent of the world's lakes can be found in North America, Asia, and Africa. But lakes are also found in frigid Antarctica, where the surfaces of

Many animals depend upon lakes for survival.

Lakes Bonney and Vanda stay frozen all year long. In dry Australia, Lake Eyre covers 3,400 square miles (8,900 sq km) when seasonal rains fall, only to dry up when the rains cease.

What is a lake? **Limnology**, the scientific study of lakes, gives us a simple definition: a slow-moving or standing body of water surrounded completely or nearly completely by land. Basically, lakes are collection points for incoming water—whether that water comes from rain, rivers, or melting snow. Some experts say the term "lake" applies to bodies of water with a surface area larger than one square acre (.4 sq h). A smaller lake may be called a pond, and a very large lake may be called a **sea**. In fact, the largest lake in the world is the Caspian Sea, located between Asia and Europe. The Caspian Sea covers 143,630 square miles (373,438 sq km).

Most lakes are **freshwater** lakes, meaning that the water contains very little salt. Freshwater lakes hold 98 percent of the world's usable surface water. A few lakes, however, are

Melting mountain snow often feeds lakes located in higher elevations, such as Jackson Lake in Wyoming (top). But no matter where their water comes from, all healthy lakes support a variety of aquatic life.

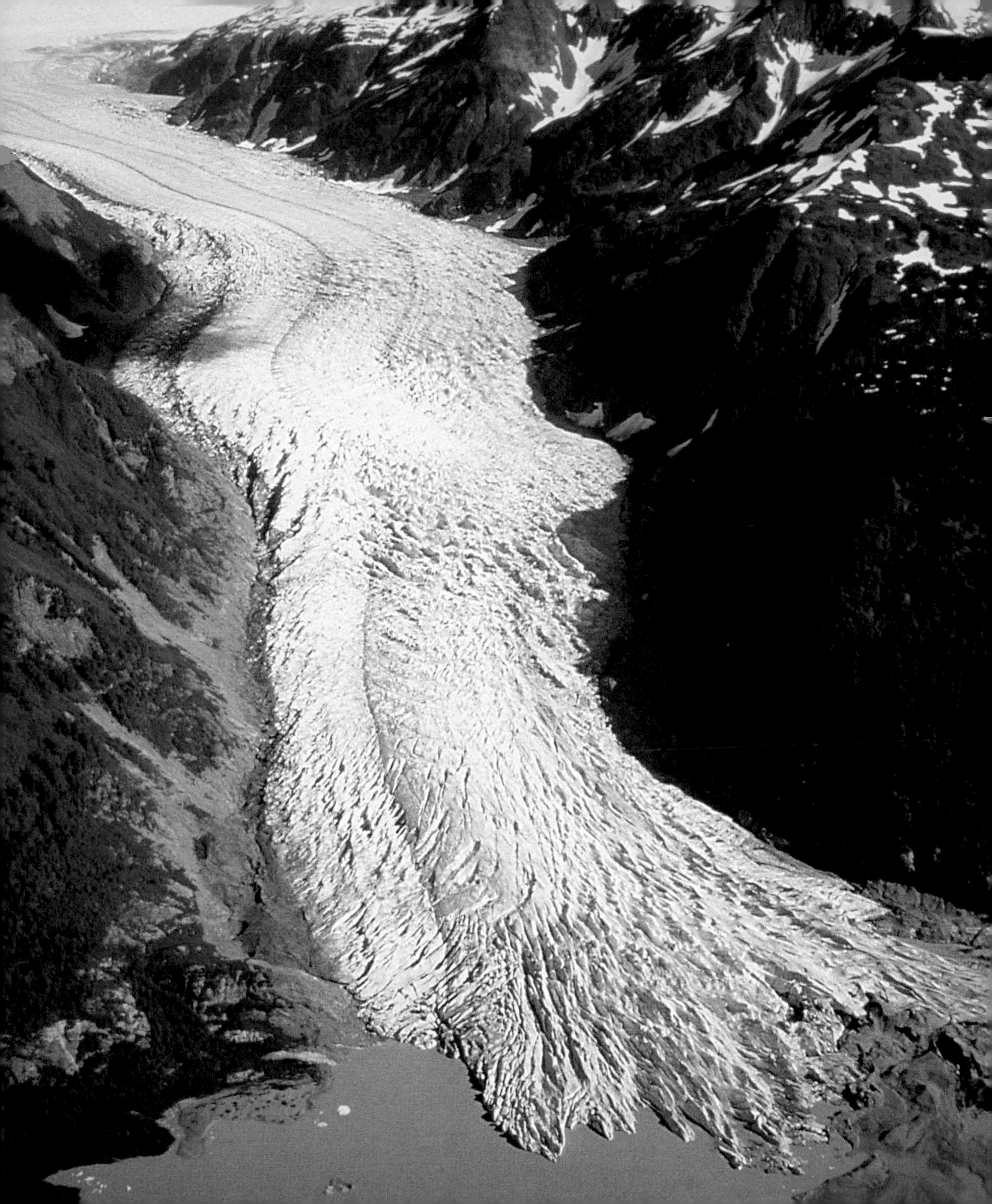

saltier than the ocean. The Great Salt Lake in Utah has water that is eight times saltier than ocean water. When the lake was formed, water carried salt from the surrounding rock formations into the lake. Then, climatic changes robbed the area of the rains that had once provided it with freshwater. Gradually, as the sun evaporated much of the water, the lake became saltier and saltier.

Just as lakes vary in their location, size, and water type, the forces that made them also vary. **Glacier** movement, powerful shifts in the earth's crust, and human activity are among the many ways in which lakes have been formed. These and other factors created **basins**, or lake beds—depressions that filled with water and became lakes.

Most lakes were created by glacier movement. Glaciers are huge sheets of **ice** formed from snow. At various times in Earth's history, temperatures were very cold and many glaciers became so massive that their own weight caused them to move, scraping the earth's surface like giant, cold bulldozers. When the climate became warmer, the glaciers melted and

During prehistoric times, huge glaciers once covered much of the northern half of the world. Their advancing and retreating motion, caused by global temperature changes, shaped the land and created many lake basins.

retreated, leaving behind depressions and valleys that became lake basins. Often a glacier deposited debris at its edge, just as a bulldozer does when it stops pushing soil.

This debris, called a **moraine**, blocked the ends of valleys carved by the glaciers, allowing the depressions to fill with water and become lakes. The Finger Lakes in New York were formed in this way. Other times, glaciers simply dug out holes when they pushed across flat regions. Many of the small lakes in the northern United States and southern Canada were formed in depressions left behind by glaciers. So were the five **Great Lakes**, located along the border between the United States and Canada. These amazing lakes hold 20 percent of the world's freshwater and cover more than 100,000 square miles (260,000 sq km).

Other lakes were formed by gradual movements in the earth's crust. Lake Okeechobee in Florida, for instance, was formed when a depression of the ocean floor was pushed

No topography was strong enough to withstand a glacier's slow, steady push—not even rugged mountain ranges. The holes, or depressions, that glaciers carved out of the earth often filled with water and became lakes.

upward and out of the ocean. Lake Baikal in Russia, on the other hand, was formed when sections of land sank away. With an average depth of 2,428 feet (740 m), Lake Baikal is the world's deepest lake.

Even humans make lakes. By damming rivers, people create lakes called **reservoirs**. One of the oldest reservoirs is Lake Parakrama Samudra in Sri Lanka, thought to be 1,600 years old. Its water is used to irrigate crops, but there are many other reasons to dam a river and create a lake. A dam prevents flooding during high-water periods, while water released from a reservoir powers turbines to produce electricity. In many cases, a reservoir means increased recreational uses, such as boating and fishing, that may attract tourists (and revenue) to the area.

Once created, lakes offer life to all types of creatures. Together, a lake's plants, animals, and insects make up its **biotic community**. Biotic communities, no matter how large, are only as healthy as their smallest members—tiny plants and animals called plankton. Plankton plants absorb **energy**

Reservoir water can be used to generate hydroelectric power.

from the sun and, when eaten, make that energy available to other creatures within a lake. The transfer of energy as one creature eats another is part of what is called the **food chain**.

While the food chain begins with the smallest of creatures, it can end with a large, powerful fish such as the muskellunge, nicknamed "the **muskie**." Called the "tigers of freshwater," muskies roam waters in the northern United States and Canada and can weigh up to 70 pounds (32 kg). Between the plankton and the muskie is a bridge built of many small creatures. For example, insects that feed on plankton are eaten by small fish called minnows. Minnows are eaten by larger fish called perch, which in turn may be consumed by muskies.

A person fishing might then catch a muskie and eat it, taking in the energy that began with the sun and the plankton. Or the muskie may resist being caught and die from other causes. Its body would settle on the bottom of the lake, where **bacteria** and fungi would break down the energy held in the

The sunfish is a relatively small freshwater fish, but it plays an important part in a lake's food chain. It keeps populations of smaller creatures under control and provides larger fish with the food they need to survive.

muskie's body into food forms that could be consumed by the plankton plants and animals.

Not only do lake creatures depend on each other, they also depend on the lake itself. In areas where **temperatures** vary greatly from season to season, lakes undergo amazing changes. It is these changes that prevent lakes as far north as the Arctic Circle from freezing completely during months of below-freezing temperatures. Freshwater is heaviest at about 40 °F (4 °C). Once ice covers a lake, the coldest water will remain just under the ice, while the heavier water sinks to the bottom. The ice insulates the lake and prevents the wind from making **waves**. Without waves to stir the lake, the water at the lake bottom stays at around 40 °F (4 °C). That is where many fish spend their winters.

As temperatures warm and the ice melts, waves begin to do their work. From the lake's surface, waves **circulate** warm water into the cooler depths. From the lake bottom, wave action push-

A layer of ice covers most northern lakes in the winter. Depending on a variety of factors, ice thickness may range from a thin skin to three feet (.9 m) or more.

es nutrients up and into shallow water, where they become food for bacteria, fungi, and plankton.

But as summer heats to its peak, another important change takes place within the lake. Water near the surface becomes much warmer and lighter than the water near the bottom, even with wave action. Soon there is a division based on water temperature. Between the two regions lies an area of temperature transition called the **thermocline**. The water above the thermocline is warmer and oxygen-rich, while the water below is cooler and oxygen-poor. Oxygen can move into but not down through the thermocline. Some fish, such as the **walleye**, need both cool water and a lot of oxygen to survive, so they spend their summers swimming in the thermocline.

Just as every creature living in a lake eventually dies, so, too, does the lake itself. When compared to other geological features such as mountains, lakes have a short life, from 1,000

In the spring, most lakes' ice layer melts. The cold water then sinks, displacing the warm water at the bottom and pushing it to the surface. This natural circulation of lake water is called turnover.

to 10,000 years. The Great Lakes are estimated to be 8,000 years old. Most people do not live long enough to see a lake actually die, but the disappearance of some lakes over a generation or two has been recorded.

Sometimes a lake will die when its water level becomes too high for its basin. Water spills over the edge and erodes the side of the lake bed. If the **erosion** becomes too great, the lake drains of water. But there is a more common way for a lake to die. From the time a lake is formed, it collects all sorts of material within its shores. A river may carry tiny pieces of dirt into the lake. Trees drop their leaves into the water. Plants and fish in the lake drift to the bottom after they die.

Once this material settles, it is called **sediment**, and it begins to fill the lake. Sedimentation causes the bottom of a lake to rise and its edges to creep toward the center. The lake becomes shallower and its shores expand. The lake turns into a **pond**. As the process continues, the pond becomes a marsh. Eventually, the marsh disappears, and a forest, prairie, or even a cornfield may grow where a lake once was.

Although they may seem timeless, lakes do eventually die. Pollution, sedimentation, and even an overabundance of aquatic plants such as duckweed or algae may all shorten a lake's life.

The aging of lakes is natural, but human activity often hastens the death of many lakes. Lakes provide water for drinking, irrigating crops, and producing electricity. Factories use lake water to keep machines cool, while cities and towns use lake water for the removal of municipal waste. Unfortunately, all of these processes have the potential to hurt lakes and the variety of life within them.

Sometimes people carelessly dump poisons and pollutants directly into lakes. Other times water **pollution** occurs unintentionally, but the effects can be just as damaging. For example, water draining from farm fields often contains nutrients from fertilizer that is used to make crops grow better. If a lake absorbs too many of these nutrients, it can become overly fertile and foster the growth of tiny blue-green plants called **algae**.

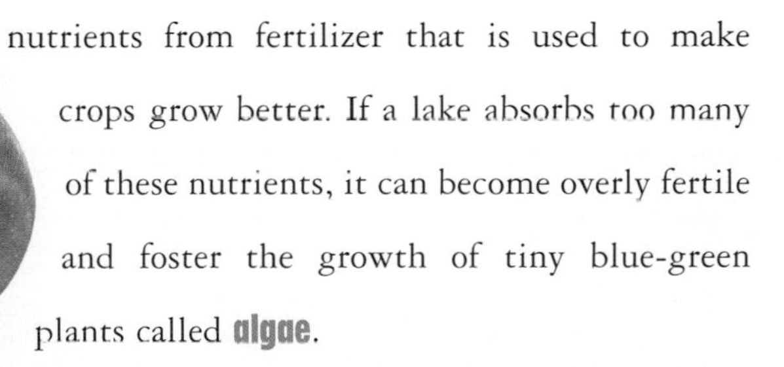

Too much algae makes the water cloudy and prevents other plants from growing. After the algae die, bacteria start the massive job of decomposition, which uses up

Many cities around the world rely on lake water to conduct their daily activities. But care must be taken to ensure that any pollution from those activities doesn't harm the lake ecosystem.

oxygen other creatures need. Such a lake is described by scientists as being eutrophic. Without high oxygen levels, fish and plants begin to die. The amount of sediment increases. The lake ages and dies more quickly than it should.

As we learn more about how our actions can hurt lakes, we can also learn how to reduce and repair the harm we do. Farmers near lakes can test their soil to determine just how much fertilizer is needed, reducing the risk of nutrient runoff to the water. Lakeshore homeowners can eliminate lawn chemicals. Stricter rules for factory waste removal can lessen the damage caused by industrial pollution. Even small actions, such as picking up litter after a picnic or fishing trip, can go a long way toward **preserving** a lake for the enjoyment and benefit of future generations.

Where lakes are found, **life** is found. We all must do our part to ensure that lakes will always be beautiful, respected parts of our world.

Our relationship with lakes should be one of respect.

ACTIVITY ONE

LAKE TURNOVER

Most lakes in the northern hemisphere undergo a great change in the fall and spring. Their water circulates, or stirs itself up, through an important natural process called turnover. This simple activity will show you how the process works.

You Will Need

- A clear glass or plastic two-quart (2 l) pitcher
- About 1/4 cup (70 ml) lake sediment (clay, silt mixture) or soil
- Tap water
- A measuring cup
- Ice water
- Blue food coloring

Preparing the Water

1. Pour the sediment into the pitcher and add tap water until the pitcher is about three-quarters full. Let the pitcher stand undisturbed for 24 hours to allow the sediment to settle. Complete the following steps the next day.
2. In the measuring cup, add drops of food coloring to about two ounces (60 ml) of ice water until the water is deeply colored.
3. Carefully pour the colored ice water against the inside of the pitcher.

Observation

The blue-colored water sank to the bottom of the pitcher because cold water weighs more than warm water. When it reached the bottom, it displaced the warm water, which then rose to the top, along with particles of sediment.

In a lake, as in any body of water, the highest concentration of oxygen is at the surface. Oxygen enters the water wherever the water and air meet. Turnovers circulate the water in a lake, ensuring that oxygen reaches all of a lake's layers, not just the top. Fall turnovers occur when autumn winds cool the surface of a lake. The cold, oxygen-rich water sinks and supplies plants and animals living on the bottom with the oxygen they need to survive. But it's not just bottom-dwelling organisms that benefit from turnovers. As the cold water sinks, the displaced warm water rises and carries nutrient-rich sediment (dead plant and animal life, mineral-laden soil) from the bottom to organisms living near the lake's surface. Eventually, this warm water too is cooled by the wind, and the cycle continues until winter freezes the surface.

During the winter, a layer of ice insulates the lake, keeping water temperatures below it warm enough to sustain most aquatic life. In the spring, the ice melts. This top layer of water is colder than the layers below it, so it sinks, thus triggering the start of spring turnover.

A C T I V I T Y

T W O

STORING HEAT

If you live near a lake, you've probably noticed that the air is just a bit warmer around the water than it is farther away. When winter arrives in northern parts of the world, lakes may not freeze over completely for a month or more after the first hard freeze or snowfall. The reason for this is that water stores heat better than land does.

To prove this, fill two identical metal cans three-quarters full, one with soil and one with water. Place both cans in a warm oven (less than 200 °F (93 °C)) for two hours. Then, carefully remove the cans and place them on a cooling rack. Using a meat thermometer, record and compare the temperature of each can every hour.

You'll find that the water retains heat better and stays warmer longer than the soil. Because of this fact, many farmers who raise cold-sensitive crops such as citrus fruit like to plant near lakes. The air rising off the lakes keeps temperatures a few degrees warmer than in locations farther inland. Oceans also retain heat well. People who live along coastlines benefit from the water's warming effect and enjoy more moderate seasons.

LEARN MORE ABOUT LAKES

Crater Lake National Park
P.O. Box 7
Crater Lake, OR 97604
http://www.nps.gov/crla

Environment Canada: Our Great Lakes
(online resource for information about Canada's Great Lakes region)
http://www.on.ec.gc.ca/glimr/intro.html

Great Lakes Information Network
c/o Great Lakes Commission
400 Fourth Street, Argus II Building
Ann Arbor, MI 48103
http://www.great-lakes.net

North American Lake Management Society
P.O. Box 5443
Madison, WI 53705
http://www.nalms.org

U.S. Environmental Protection Agency
Great Lakes National Programs Office
77 W Jackson Boulevard
Chicago, IL 60604
http://www.epa/gov/glnpo

U.S. Geological Survey's Water Resources Division
(offices in every state; check the website or your local telephone directory for listings)
http://wwwga.usgs.gov/edu

INDEX

Wherever healthy lakes are found, life is found.